Origami Dots

Andy Parkinson

About the Author

Andy Parkinson is the mathematics adviser on the island of Jersey in the Channel Islands. He has taught mathematics for over 30 years both in the UK and around the world. He sees mathematics as a creative activity, full of surprising connections. The more you look, the more you find, the more you want to look. He posts about mathematics on twitter @andyparkinson31.

For Debby

ISBN (book): 978-1-907550-19-5
ISBN (ebook): 978-1-913565-38-1

Printed and designed in the UK

Published by Tarquin
Suite 74, 17 Holywell Hill
St Albans AL1 1DT
United Kingdom

info@tarquingroup.com
www.tarquingroup.com

Introduction

Origami Dots is a collection of intriguing mathematical challenges that come from folding paper.

The black dot at the corner (vertex) of the shape is folded to the target black dot somewhere in the puzzle. The two red dots define the crease line needed for the fold.

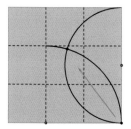 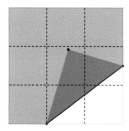

The question is always "What is the area of the paper that has been folded over?"

All the challenges are based on a grid of 144 units of squares or triangles, each with a side length of 1cm.

 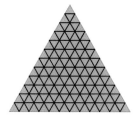

You can download blank PDF templates of the grids free of charge at www.tarquinselect. com - search for Origami Dots Grids.

In the answers, the folded area is given as a number of the 144 units of the original shape. Sections I, III and IV are based on squares and section II on triangles.

Start with an estimate, an intelligent guess. It is always interesting to find out how close your guess is to the answer. Explore by actually folding the corners of the book. What can be deduced and reasoned geometrically from the fold?

The area to be calculated in each Origami Dots puzzle can be physically made in either of two ways:

1. Cut along the edges highlighted with a scissors icon so that the black corner dot(s) can be folded onto the black target dot with the fold line(s) going through two red dots.
2. The folded area can be drawn on the diagram using the following rules:
   a. Sections I and II
      Draw lines from the red dots to the black target dot and between the two red dots.

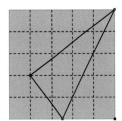
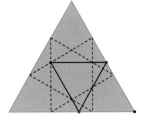

   b. Section III
      Draw a line from the red dot on the bottom line to the black target dot. Draw a 12cm line at 90° to the first line and through the top line of the puzzle. Connect the end of the 12cm line to the other red dot. Join the two red dots.

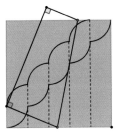

c. Section IV
   Draw two right-angled triangles each with two red dots and the black target dot
   for vertices (corners). Find the total area of the two triangles. If these triangles
   overlap then find the area of the quadrilateral whose vertices are: the black target
   dot, the top and bottom red dots and the intersection of the lines between two red
   dots (the hypotenuse of each right-angled triangle).

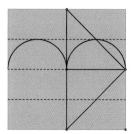 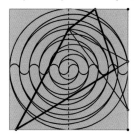

All the challenges define the target dot using the same two rules. These are –

1. Dotted grid lines are equally spaced along an edge or around a vertex or point on the
   square or triangle.
2. Arcs are part of either a circle, semicircle or quadrant.

The magnifying symbol ⚲ near the dot in question 33 highlights that the dot is very close
to, but not on, the gridline. The dot is on the outer square.

The challenge in Origami Dots is that a small change in the position of the target dot can
drastically change the logic needed to find the area of the folded section. However, there
will usually be a range of strategies possible to discover the location of any one target dot,
so consider solving the puzzle in more than one way.

Reflecting on how a puzzle was solved often provides additional insights that can help on
the challenges that follow. Also, being able to arrive at the same answer in more than one

way can give confidence that the answer is correct. Is one method more efficient or clearer to understand?

Ask (and try to answer) your own questions along the way. Learn new techniques. The book encourages curiosity. A "find out what you can" discovery approach creates interesting and varied strategies towards the solutions and mathematical skills are developed. The further we look, the more we realise there is to find.

To help the reader, each puzzle has signposts towards the techniques that might produce a possible solution. The signposts are not suggesting any "best" solution, just a hint at a possible strategy. Other insightful solutions may well exist just waiting to be discovered.

List of signposts

$\dfrac{x}{}$	Linear equations	◿	Triangles
$\lfloor x \rfloor$	Quadratic equations	◗	Arcs and Circles
⋈	Similar triangles	▱	Quadrilaterals
⬠	Pythagoras' Theorem	$\dfrac{?}{?}$	Fractions
△	Trigonometry	$\sqrt{?}$	Surds

However, if at any point the challenge becomes too great, (approximate) answers can always be found by measuring the side lengths (in centimetres) of the folded area and using a formula to calculate the area. Values for this approach will be given in the solutions.

Where answers are not a whole number, the exact answer is given either as a fraction or as a surd. Two approximations are also given (the first the exact value rounded to one decimal place and the second value, an answer that could possibly be obtained through measurement).

In the solutions, if a particular technique for solving a challenge has been expanded upon in an earlier question it may not be repeated again in the later question but left to the reader to work through. Therefore, try to work through the puzzles in the given order.

In the solution for a puzzle, the aim has been to highlight connections between puzzles and other mathematical ideas that come from these simple folds, as well as asking further probing questions for the interested reader.

For teachers, this book provides a rich source of challenges that have a similar appearance, yet hide an obvious method of solution. This requires students to choose how to approach each individual problem. Mathematical techniques are repeated in subsequent challenges, allowing opportunities to embed prior learning.

The book is divided into four sections, each with problems followed by solutions.

SECTION 1

Folding triangles on squares   1

SECTION 2

Folding triangles on triangles   31

SECTION 3

Folding trapezia   61

SECTION 4

Two folds   91

Folding triangles on squares

In the middle

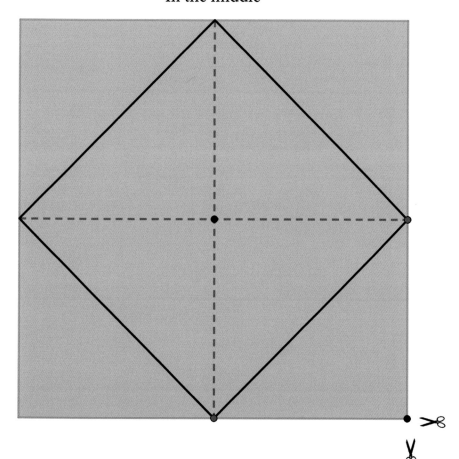

2

Going north-west

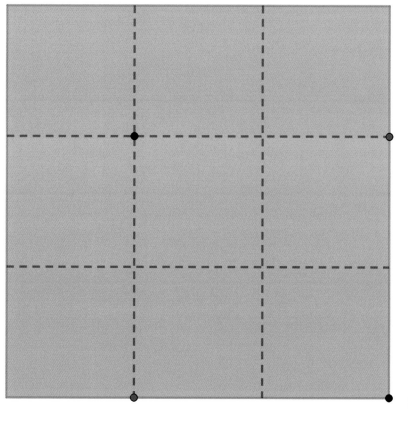

3

Two arcs are better than one

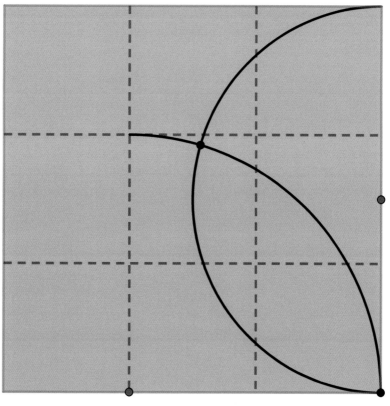

4

Tilted Square

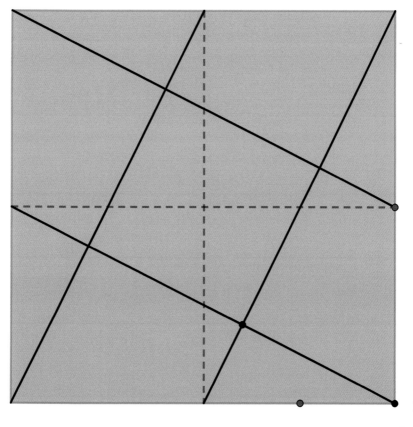

5×5

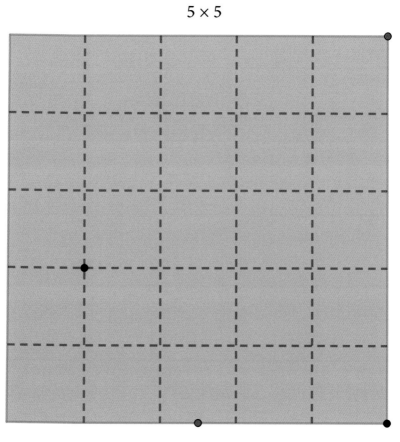

Trisection?

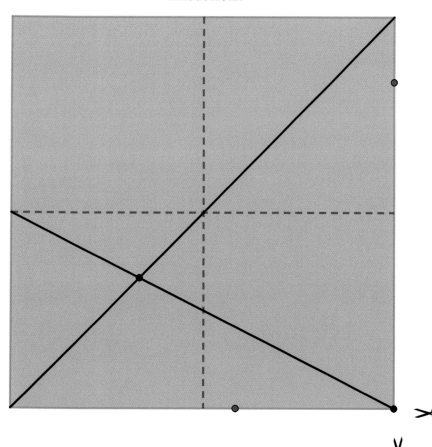

7

Star

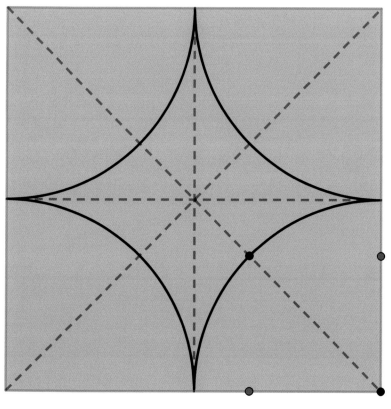

Criss-cross

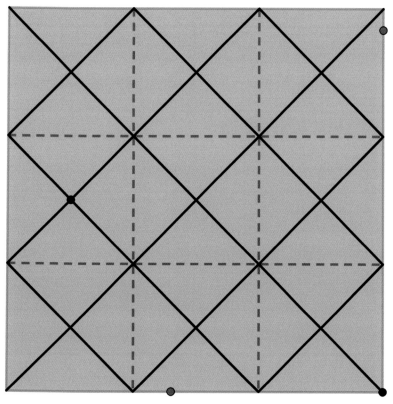

9

Edge of a flower

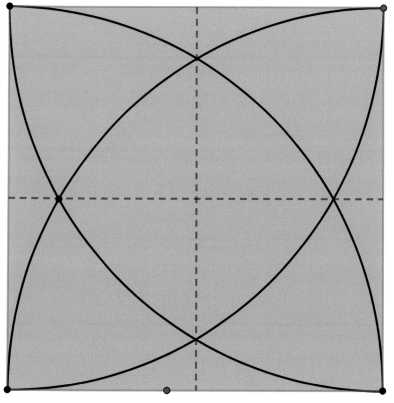

10

Midpoint of an arc

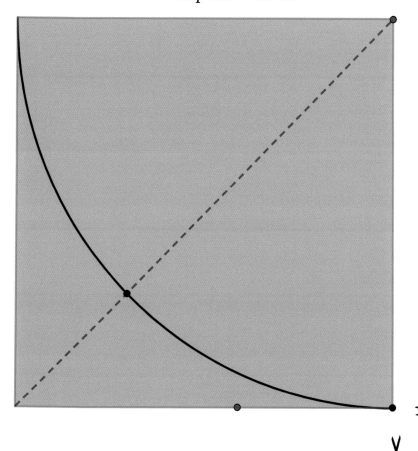

11

Cutting a leaf in two

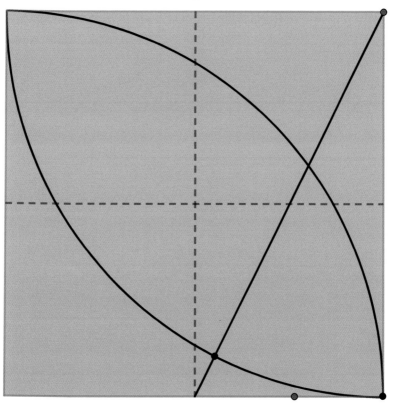

Solutions and Answers

Answers   (Let A be the unit area of the folded region)

1

Folded area, A = 18 squares

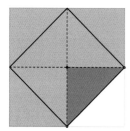

As the dotted lines bisect each side, the target dot is in the centre of the square. Therefore ($\therefore$), using the area of the triangle

$$A = \frac{1}{2} \times 6 \times 6$$
$$A = 18$$

OR the folded area is one eighth of the square.

$$\therefore A = \frac{1}{8} \times 144$$
$$A = 18$$

Aside: If we were to calculate the uncovered area it would be area of the square – twice the folded area = $144 - 2 \times 18 = 144 - 36$ and not 142×18.

Multiplying before subtracting The order of operations makes sense from the fold doubling the thickness of the paper, as opposed to calculating the expression from left to right.

2

Folded area, A = 32 squares

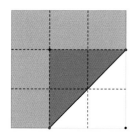

As the corner of the square is being folded onto the diagonal of the square, the folded triangle is isosceles.

$$= \frac{2}{3} \times 12 = 8$$
$$A = \frac{1}{2} \times 8 \times 8$$
$$A = 32$$

OR As 5 of the 9 divided squares of the original square remain uncovered

$$A = \frac{1}{2} \times \frac{4}{9} \times 144$$
$$A = \frac{2}{9} \times 144$$
$$A = 32$$

3

Folded area, A = 24 squares

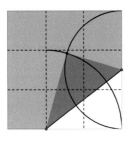

As the target dot is at the intersection of both arcs, the radius of each arc is the side length of a side of the triangle.

$$A = \frac{1}{2} \times 6 \times 8$$

$$A = 24$$

$$= \frac{1}{6} \times 144$$

The folded triangle has sides 6, 8 and 10. An enlargement of (similar to) the right angled triangle with sides 3, 4 and 5, which will feature repeatedly throughout the book.

4

Folded area, A = 9 squares

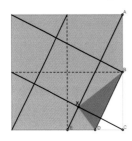

Let $\overset{\frown}{EAC} = \theta = \overset{\frown}{XCE}$ *(by rotational symmetry),*

$\overset{\frown}{CEA} = \mu,\ \theta + \mu = 90°$

$\overset{\frown}{DXC} = \overset{\frown}{XCD} = \theta$ *from the fold*

$\therefore \overset{\frown}{EXD} = \mu$

$\therefore \Delta XDE$ *is isosceles* $XD = ED$,

$XD = CD$ *from fold*

$\therefore ED = CD = \frac{1}{2}$ *of* $6 = 3 = XD$

$\therefore A = \frac{1}{2} \times 3 \times 6 = 9\ BX = BC$

OR $\Delta BCD = \frac{1}{4}$ *of* $\frac{1}{4}$ *of* $144 = 9$

Aside: when the two left hand side vertices of the square form a triangle with D, what do you notice?

5

Folded area, A = 36 squares

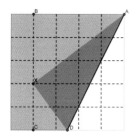

$\triangle AXB$ is similiar to ($\sim$)$\triangle XDC$

$AB = 2XC \therefore AX = 2XD$

$XD = 6$

$A = \dfrac{1}{2} \times 6 \times 12 = 36$

OR As $XD = 6$, D is the midpoint of the side

$\therefore A = \dfrac{1}{4}$ of $144 = 36$

Aside, looking at dotted grid squares in $\triangle AXB$ we again see lengths 3, 4 and 5. Alternatively, if X is moved 1 grid square to the right and 1 square down, how does the area change? A similar logic can be used and, as with any of the other puzzles, this is a good example of slightly altering the setup to form a new question. $\left(A = 24, \dfrac{1}{6} \text{ of } 144 \right)$

6

Folded area, A = 25 squares

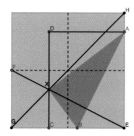

As $\triangle EXH \sim \triangle FXG$ and $EH = 2FG$, $CE = 2CG$.

$\therefore CG = \dfrac{1}{3} GE = 4$, hence the trisection suggested in the title

Using Pythagoras' Theorem on $\triangle BXC$ where $XB = l$, and $BC = 8 - l$ as $BC = CE - BE$ and $BE = XB$ from the fold

$4^2 + (8 - l)^2 = l^2$

$16 + 64 - 16l + l^2 = l^2$

$16l = 80$

$l = 5 \quad \therefore BX = 5$ and $BC = 3$

This last technique is worth studying as it reoccurs in a number of the later puzzles

$\triangle DAX \sim \triangle CXB$ and has sides 6, 8, 10

$A = \dfrac{1}{2} \times 5 \times 10 = 25$

OR   Using Pythagoras' Theorem $EX = \sqrt{4^2 + 8^2} = \sqrt{80} = \sqrt{16} \times \sqrt{5} = 4\sqrt{5}$ (leaving the answer in this form is called a surd).

$$AB = \sqrt{5^2 + 10^2} = \sqrt{125} = 5\sqrt{5}$$

$$A = \frac{1}{2}\ \textbf{\textit{area of}}\ AXBE = \frac{1}{2} \times \left(\frac{1}{2} \times 4\sqrt{5} \times 5\sqrt{5}\right) = 5\sqrt{25} = 5 \times 5 = 25$$

7

Folded area, A = 9 squares

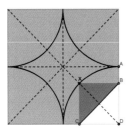

$DX = DA = 6$

Let $BX = CX = l$

Using Pythagoras' Theorem on $\triangle CDX$, $l^2 + l^2 = 6^2$

$2l^2 = 36, \qquad l^2 = 18$

$A = \frac{1}{2}\,l^2 = 9$ By measurement, folded area $\approx \frac{1}{2} \times 4.2 \times 4.2 = 8.82 = 8.8$ *(1 d. p.)*

As in question 4 can you find other folds with an area of 9 squares?

8

Folded area, $A = \dfrac{578}{15} = 38.533333\ldots\ldots$ The digit three infinitely repeats.

$= 38.5\dot{3} = 38.5$ squares *(to 1 decimal place).*

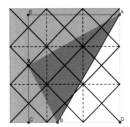

Using Pythagoras' Theorem on $\triangle BXC$, $6^2 + (10 - l)^2 = l^2$

$l = \dfrac{34}{5} = 6.8$ Let $EX = m$ As $\triangle BCX \sim \triangle XEA$, $EA = \dfrac{10}{6}\ CX\ \therefore m = \dfrac{10}{6} \times 3.2$

$m = \dfrac{16}{3}, \therefore AX = AD = \dfrac{16}{3} + 6 = \dfrac{34}{3}$

$A = \dfrac{1}{2} \times \dfrac{34}{5} \times \dfrac{34}{3} = \dfrac{578}{15}$

By measurement $A \approx \dfrac{1}{2} \times 6.8 \times 11.3 = 38.4 = 38.4$ *(1 d.p.)*

9

Folded area $= A = 24\sqrt{3} = 41.569219\ldots = 41.6$ squares. (1 *d. p.*) Unlike in a fraction, in the decimal form of a surd the pattern of digits never repeats.

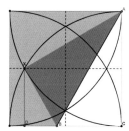

$AX = XC = 12$

$\therefore \Delta AXB$ is equilateral

When the fold is made it halves the equilateral triangle into two right angled triangles with angles 30°, 60° and 90° where the longest side is always twice the shortest side

$\widehat{CBA} = \widehat{ABX} = \widehat{XBD} = 60°$

$\Delta DXB \sim \Delta XAB \therefore XB = 2DB$

Using Pythagoras' Theorem on

$\Delta DXB, \; 6^2 + l^2 = (2l)^2$

$3l^2 = 36$

$l = \sqrt{12} = 2\sqrt{3}$

$A = \dfrac{1}{2} \times 12 \times 4\sqrt{3} = 24\sqrt{3}$

By measurement, $A = \dfrac{1}{2} \times 12 \times 6.9 = 41.4$

10

Folded area, $A = 72\sqrt{2} - 72 = 29.823376\ldots = 29.8$ *(1 d. p.)*

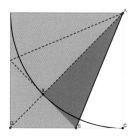

The answer is not a calculation waiting to be worked out. It is like 110 with two parts - one hundred and (one) ten, simply best written in two parts. 72 root 2's less 72.

The decimal form helps compare the area with other challenges.

$A = \dfrac{1}{2} \times 12 \times l = 6l$

From the diagram, $4A + \dfrac{1}{2}(12-l)^2 = 144$

$24l + \dfrac{1}{2}\left(144 - 24l + l^2\right) = 144$

$48l + 144 - 24l + l^2 = 144 \times 2$

$144 + 24l + l^2 = 144 \times 2$

$(12 + l)^2 = 144 \times 2$

$l = 12\sqrt{2} - 12$

$A = 6l = 72\sqrt{2} - 72$

By measurement, $A = \dfrac{1}{2} \times 5 \times 12 = 30$

OR as the length of a square's diagonal is $\sqrt{2}$ times it side length, as in $\triangle XDB$

$$l + l\sqrt{2} = 12$$

$$l\left(1 + \sqrt{2}\right) = 12$$

$$l = \frac{12}{1 + \sqrt{2}}$$

$$l = \frac{12}{1 + \sqrt{2}} \times \frac{\sqrt{2} - 1}{\sqrt{2} - 1}$$

$$l = \frac{12\left(\sqrt{2} - 1\right)}{\left(\sqrt{2}\right)^2 - 1^2}$$

$$l = 12\sqrt{2} - 12$$

The highlighted rectangle is in the same ratio (similar to) as a standard piece of A4 paper as

$$\left(24 - \left(24 - 12\sqrt{2}\right)\right):12 = \sqrt{2}:1$$

As $24 - 12\sqrt{2} \approx 7$

$$\sqrt{2}:1 \approx 17:12$$

$$\sqrt{2}\left(= 1.4142\right) \approx \frac{17}{12}\left(= 1.4167\right)$$

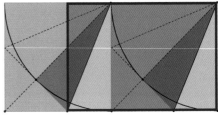

11

Folded area, $A = 72\sqrt{5} - 144 = 16.996894\ldots\ldots = 17.0$ (1 d. p.)

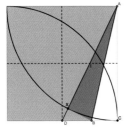

$$AD = \sqrt{6^2 + 12^2} = 6\sqrt{5}$$

$$\therefore DX = 6\sqrt{5} - 12$$

$$BX = l,\ l^2 + \left(6\sqrt{5} - 12\right)^2 = \left(6 - l\right)^2$$

$$l = 12\sqrt{5} - 24$$

$$A = \frac{1}{2} \times 12 \times \left(12\sqrt{5} - 24\right) = 72\sqrt{5} - 144$$

By measurement $\frac{1}{2} \times 12 \times 2.8 = 16.8$

OR $\widehat{BAC} = \theta$, $\tan(2\theta) = \dfrac{2\tan\theta}{1 + \tan^2\theta} = \dfrac{1}{2}$,

solving for $\tan\theta$, $\tan\theta = \sqrt{5} - 2$,

and $l = 12\tan\theta$

Aside: can you show $6\sqrt{5} - 12 \approx \sqrt{2}$? (*Hint*: Compare with the answer in question 10.)

Can you see the connection between the diagram in this question and in question 6 and the number $\sqrt{5}$? What is the significance of the two arcs and the shape of the folded area? (*Hint*: Compare the positions of the target dots in Section I with the target dots in Section III.)

Folding triangles on triangles

Folding triangles on triangles

12

Four Triangles

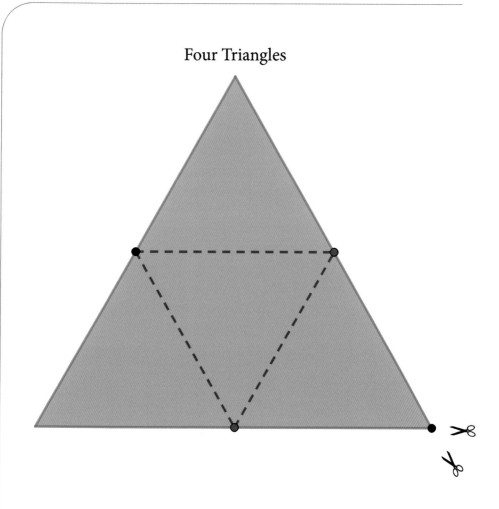

More Triangles

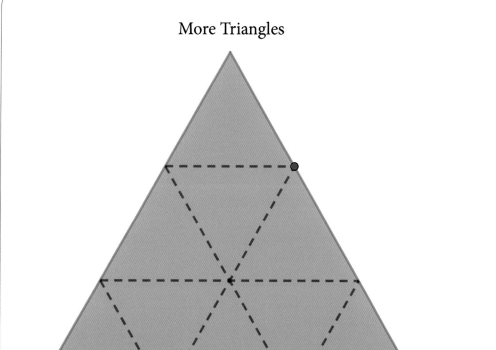

Trapped Circle

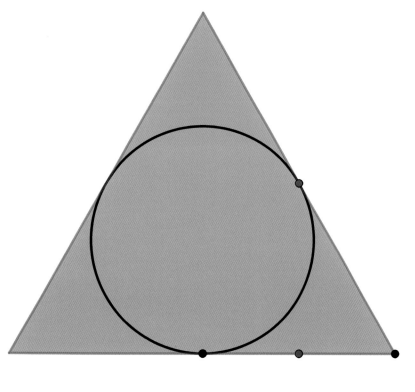

Half Cuts

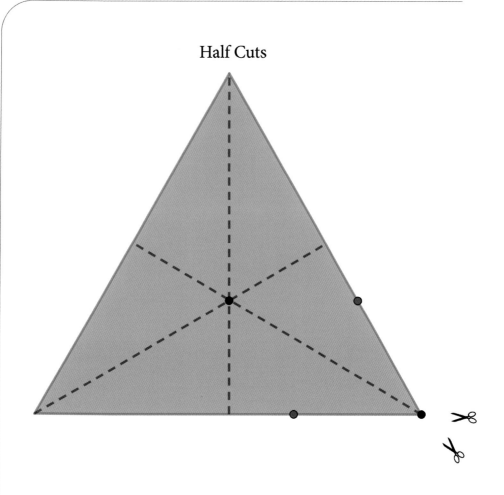

16

Six Points

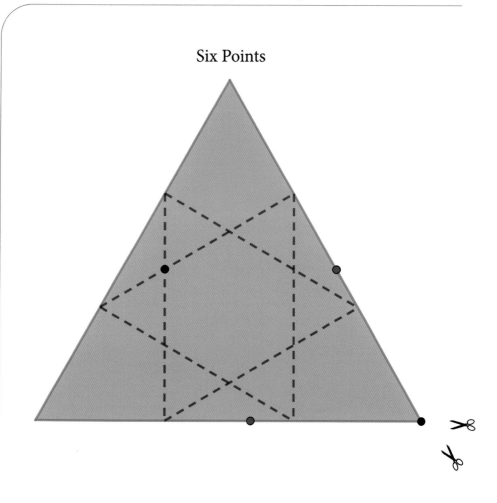

17

Overlapping Triangles

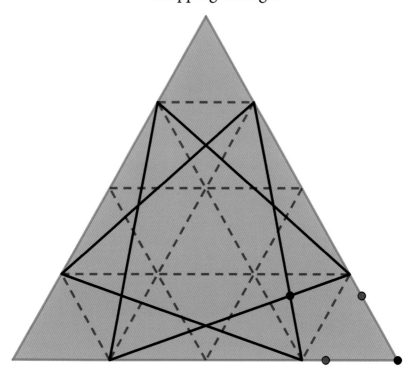

18

Criss-Cross Cube

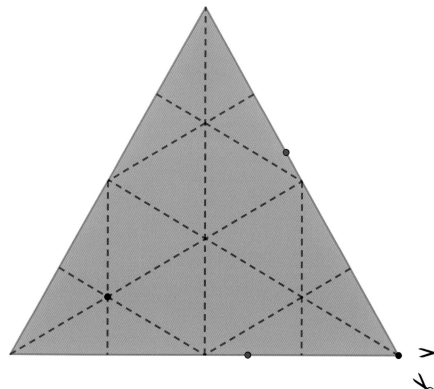

19

Many Triangles

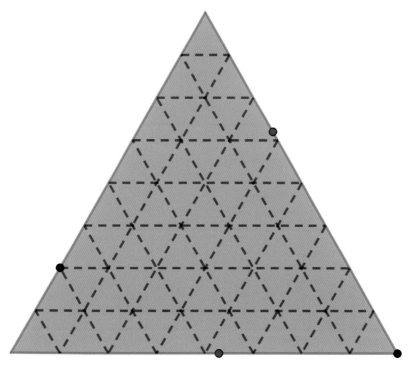

20

Mini Snooker

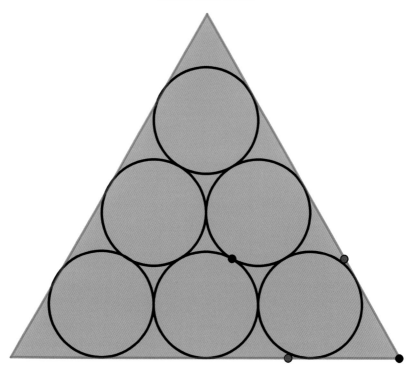

Projected Cube

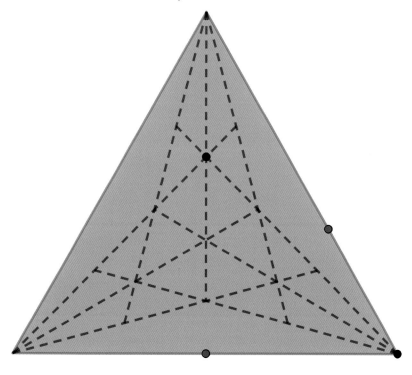

Round and Round

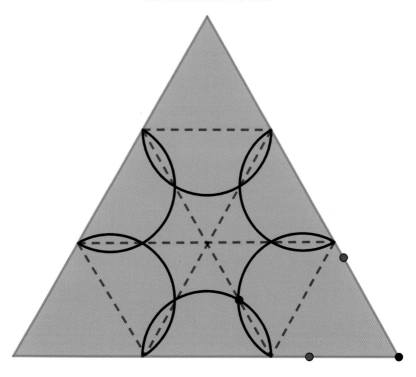

Solutions and Answers

12

Folded area, A=36 triangles

$\triangle ABX$ is $\dfrac{1}{4}$ of the large triangle

$\therefore A = \dfrac{1}{4}$ of 144

$= 36$

OR counting unit (1cm) triangles in rows triangles

$= 11 + 9 + 7 + 5 + 3 + 1$

$= 36$

Calculating the area on the square grid relies on using surds

Area in cm² $= \dfrac{1}{2} \times 6 \times 6 \times \sin\left(60°\right)$

$= 9\sqrt{3} \approx 15.6$

OR Heron's Formula

$A = \sqrt{s\left(s-a\right)\left(s-b\right)\left(s-c\right)}$ where a,b,c are side lengths

$s = \dfrac{1}{2}(a + b + c)$

$A = \sqrt{9\left(9-6\right)\left(9-6\right)\left(9-6\right)}$

$= \sqrt{243} = 9\sqrt{3}$

The answers to the remaining puzzles in this section will be expressed in unit triangles.

13

Folded area, A = 32 triangles

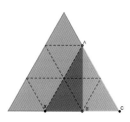

The area left uncovered is $\dfrac{5}{9}$ of the large triangle.

$\therefore$ *Folded fraction* $= \dfrac{1}{2}\left(1 - \dfrac{5}{9}\right) = \dfrac{2}{9}$

$A = \dfrac{2}{9}$ of 144

$A = 32$

14

Folded area, A = 18 triangles

By symmetry A and B are at the midpoints on the sides of the triangle.

$$\therefore A = \frac{1}{2} \times \frac{1}{4} \times 144$$

$$= 18$$

OR It can be shown that the folded area, A in triangles is

$$= AX \times BX$$

By folding, XA = CA = 6 and
XB = BC = 3

$$\therefore A = 6 \times 3 = 18 \text{ triangles}$$

NB When calculating conventional areas in squares, multiplied lengths must be at 90° to each other. Here, calculating in equilateral triangles, multiplied lengths must at 60° to each other.

15

Folded area, A = 16 triangles

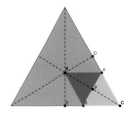

$\triangle ACG \sim \triangle AXG \sim \triangle AXD$ *(from the fold)* and

$$\triangle CXD = \frac{1}{6} \text{ large triangle}$$

$$A = 2 \times \frac{1}{3} \times \frac{1}{6} \times 144$$

$$= 16$$

OR $XG = \frac{1}{3} CE$ and $XB = XA = \frac{1}{3} \times 12 = 4$

$$A = 4 \times 4 = 16$$

Alternatively the folded area, given $\triangle ABC$ is similiar to the large triangle

$$= \left(\frac{1}{3}\right)^2 \times 144$$

$$= 16 \text{ triangles}$$

16

Folded area, $A = 28\frac{4}{9} = 28.444444\ldots = 28.\dot{4} = 28.4$ triangles $(1\,d.\,p.)$

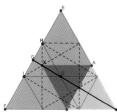

In $\triangle CEF$, $CD = \frac{2}{3}CG$

In $\triangle DHI$, $DX = \frac{2}{3}GD$

$\therefore CX = \left(\frac{2}{3} + \frac{2}{3} \times \frac{1}{3}\right)CG = \frac{8}{9}CG$

$\therefore AC = \frac{1}{2} \times \frac{8}{9}$ of CE

$= \frac{4}{9} \times 12$

$= 5\frac{1}{3} = AX = BX$

$A = 5\frac{1}{3} \times 5\frac{1}{3}$

$= 28\frac{4}{9}$

Considering dividing $\triangle CEF$ into 81 smaller triangles

$A = \frac{1}{4}(81-17) = 16$ larger triangles

Each triangle is multiplied $\left(\frac{12}{9}\right)^2$ the area of the unit triangle

$A = \left(\frac{12}{9}\right)^2 \times 16 = 28\frac{4}{9}$

By measurement $5.3 \times 5.3 = 28.09$
$ = 28.1$ (1 d.p.)

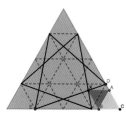

17

Folded Area $A = 5\frac{1}{16} = 5.0625 = 51$ triangles $(1\,d.\,p.)$

$CE = 9$, $CD = 3$ $\therefore CE = 3CD$

$\triangle CDE \sim \triangle BXE$ $\therefore BE = 3BX$

Let $BX = x = BC$,

$3x = 9 - x$

$x = 2\frac{1}{4}$

By symmetry

$A = 2\frac{1}{4} \times 2\frac{1}{4} = 4 + \frac{2}{4} + \frac{2}{4} + \frac{1}{16}$

$= 5\frac{1}{16} = 5.0625$

By measurement,

$2.3 \times 2.3 = 5.29 = 5.3$ $(1\,d.\,p)$

18

$A = 32\dfrac{2}{3} = 32.666666\ldots\ldots = 32.\dot{6} = 32.7$ triangles $(1\,d.\,p)$

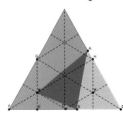

As $\triangle DGF$ is equilateral, $EX = \dfrac{1}{3}\,GE$

and $GE = \sqrt{3}\;DE$

$EX = \sqrt{3}$

Applying Pythagoras' Theorem to $\triangle BEX$,
$BX = x$

$\left(\sqrt{3}\right)^2 + \left(9-x\right)^2 = x^2$

$x = 4\dfrac{2}{3}$

Applying Pythagoras' Theorem to $\triangle AXH$,
$AH = y$

$\left(4\sqrt{3}\right)^2 + y^2 = \left(6+y\right)^2$

$y = 1$ and $AC = 7 = AX$
OR albeit less efficient.......

As $\widehat{DXH} = \widehat{BXA} = 60°$

$\widehat{HXA} = \widehat{DXB} = \theta$

Applying The sine rule to $\triangle BXD$

$\dfrac{\sin(150)}{4\dfrac{2}{3}} = \dfrac{\sin(\theta)}{\dfrac{4}{3}}$

$\sin(\theta) = \dfrac{1}{7} \quad \therefore XA = 7AH$

$XH = \dfrac{2}{3}\,FH = 4\sqrt{3}$

Applying Pythagoras' Theorem to
$\triangle AXH$, $AH = y$

$\left(4\sqrt{3}\right)^2 + y^2 = \left(7y\right)^2$

$48 = 48y^2 \therefore y = 1$ and $AC = 7 = AX$

OR comparing with other method
$6 + y = 7y \quad \therefore y = 1$

$A = 7 \times 4\dfrac{2}{3} = 32\dfrac{2}{3}$

By measurement $7 \times 4.7 = 32.9$

19

Folded Area, $A = 43\dfrac{16}{35}\;43.4\dot{5}71428\ldots\ldots = 43.5$ triangles $(1\,d.p.)$

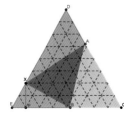

Applying Pythagoras' Theorem to $\triangle FXE$
& $\triangle BXF$ and $BX = x$

$\left(3^2 - \left(1\dfrac{1}{2}\right)^2\right) + \left(10\dfrac{1}{2} - x\right)^2 = x^2$

$x = 5\dfrac{4}{7}$

$\triangle EBX \sim \triangle DXA$ and $AX = y$

$\dfrac{y}{9} = \dfrac{\dfrac{39}{7}}{\dfrac{45}{7}}$

$$y = 7\frac{4}{5}$$

$$A = 5\frac{4}{7} \times 7\frac{4}{5} = 43\frac{16}{35}$$

By measurement

$5.6 \times 7.8 = 43.68 = 43.7 \,(1\,d.\,p.)$

Aside: The fraction of the triangle left uncovered is $\dfrac{111}{280} = 0.396428571\ldots\ldots$ uses all ten digits in the first ten of its *decimal expansion*.

20

Folded area, A=12 triangles

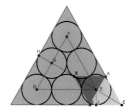

As $DG = 2r$ r *is radius of the circle*

$EG = 2\sqrt{3}r$, $EG = r$

As $\triangle EJC$ *has angles 30°, 60° and 90°* $EC = 2r$

$\therefore CH = 3r + 2\sqrt{3}r = 6\sqrt{3}$

Multiplying both sides by $\frac{2}{3}\sqrt{3}$

$2\sqrt{3}r + 4r = 12$

$\therefore 3r + (12 - 4r) = 6\sqrt{3}$

$r = 12 - 6\sqrt{3}$

$$\frac{XC}{2} = \frac{\sqrt{3}}{2}r + r = r\left(\frac{\sqrt{3}}{2} + 1\right)$$

$$= 12\left(1 - \frac{\sqrt{3}}{2}\right)\left(\frac{\sqrt{3}}{2} + 1\right)$$

$$= 12\left(1^2 - \left(\frac{\sqrt{3}}{2}\right)^2\right)$$

$$= 12 \times \frac{1}{4}$$

$$= 3 \,(NB\ this\ is \neq EC = 2r = 3.21539\ldots\ldots)$$

$$(AX)^2 = 3^2 + \left(\frac{AX}{2}\right)^2$$

$$= \frac{3}{4}AX^2 = 9$$

$A = AX^2 = 12$

by measurement $= 3.5 \times 3.5$

$= 12.25 = 2.3 \,(1\,d.\,p.)$

21

Folded area, A = $36\sqrt{3}$ – 36 = 26.353829...... = 26.4 triangles (1 d. p.)

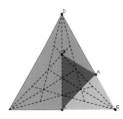

From the fold

$AX = AC \therefore \widehat{CXA} = \widehat{ACX} = 15°$ and

$\widehat{BXA} = \widehat{ACB} = 60° \therefore \widehat{BXC} = 45°$

As $XCB = 45° \widehat{CXB} = 90°$

$\therefore B$ *is the midpoint of the base*

$BX = 6$

$\widehat{AXD} = 120°$

$\widehat{BAC} = \widehat{XAB} = (180 - 45 - 60) = 75°$

$\therefore \widehat{XAC} = 150°$

$\widehat{DAX} = 30° = \widehat{XDA}$

$\therefore AX = DX = BD - BX$

$= \sqrt{12^2 - 6^2} - 6$

$= 6\sqrt{3} - 6$

$A = BX \times AX = 6\left(6\sqrt{3} - 6\right)$

$= 36\sqrt{3} - 36 = 26.3538......$

By measurement $6 \times 4.4 = 26.4$

22

Folded area, A = 9.8 triangles

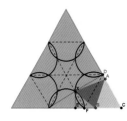

$CD = 4, DX = 2\sqrt{3}, EX = \sqrt{3}, EF = 1$

Let $BX = x$, *then* $x^2 = \left(\sqrt{3}\right)^2 + (5 - x)^2$

$x = 2.8$

Let $AD = y$, *then* $y^2 + \left(\frac{1}{2}\sqrt{8^2 - 4^2}\right)^2 = (4 - y)^2$

$y = 0.5 \therefore AC = 3.5 = AX$

$A = AX \times BX = 3.5 \times 2.8$

$= 9.8$

Folding trapezia

23

On the pitch

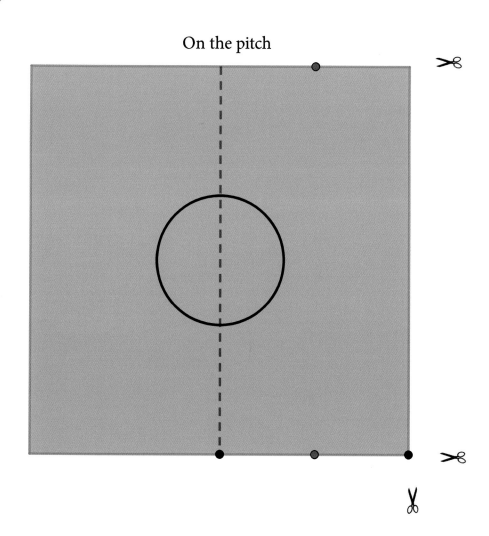

24

Playing the percentages

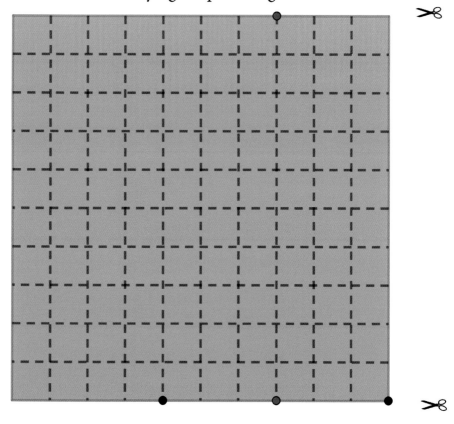

25

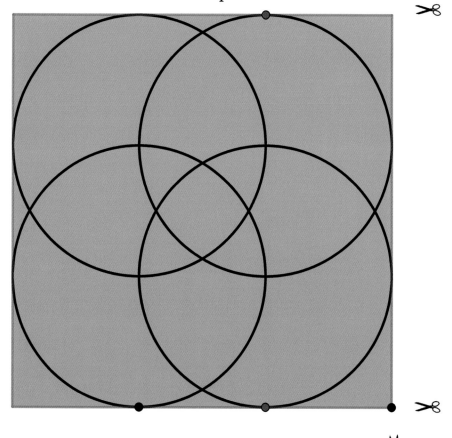

Almost a complete Venn

26

Half full or half empty?

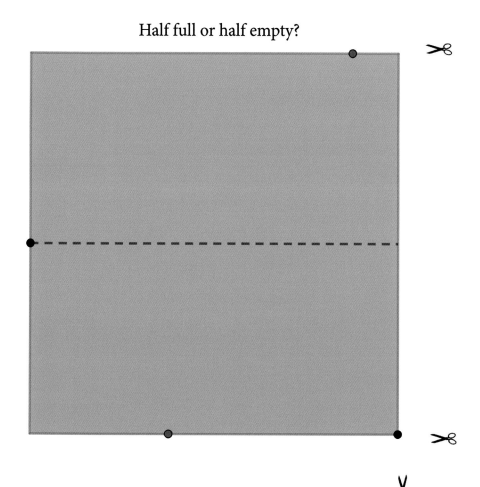

27

Ever decreasing semicircles

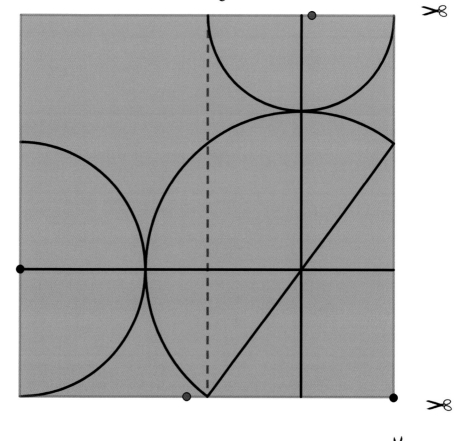

28

Circular foliage

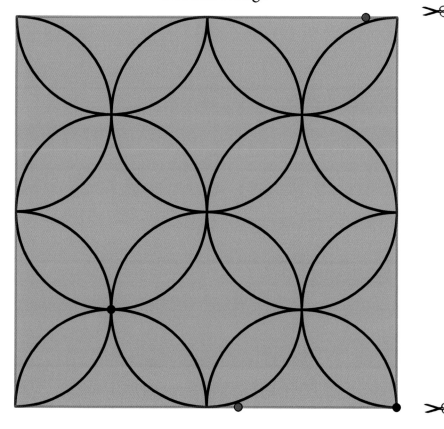

29

Quadrant twist

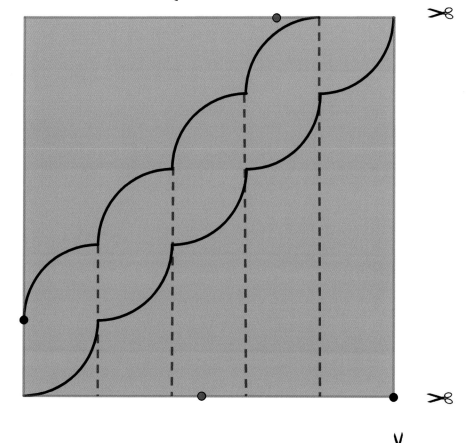

30

Eight o' clock

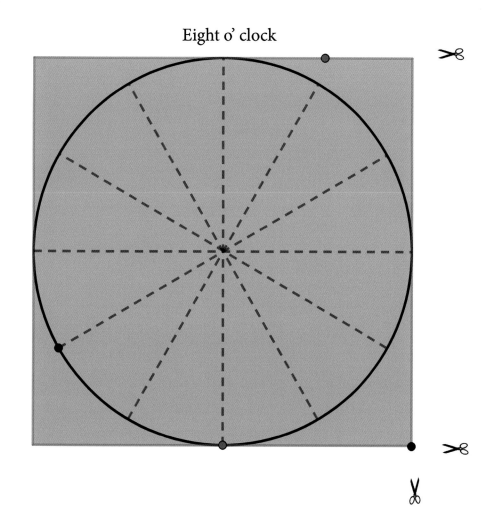

31

Quadrants great and small

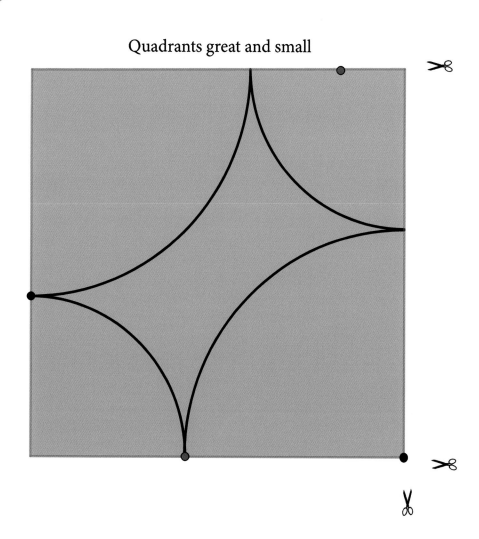

32

Links of circles

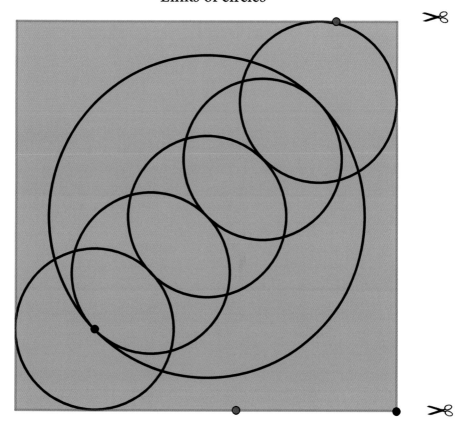

33

Pieces of one

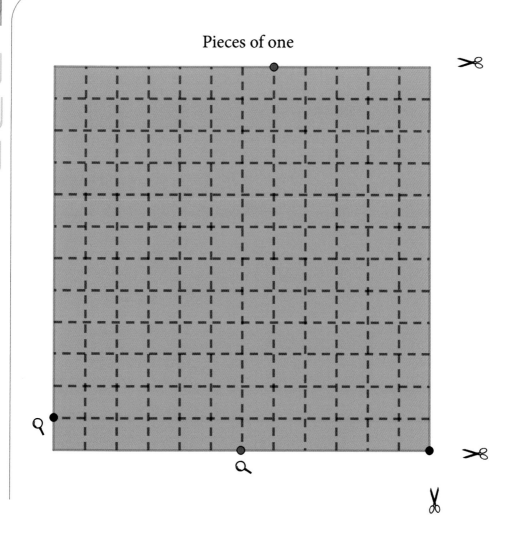

Solutions and Answers

23

Folded area, A = 36 squares

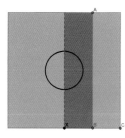

$$XB = \frac{1}{2} \text{ of } \frac{1}{2} \text{ of } 12 = 3$$

$$\therefore A = 3 \times 12$$

$$= 36$$

Aside: What is the ratio of the fraction of the folded region in the centre circle to the fraction of the unshaded region in the centre circle?

24

Folded area, A = 43.2 squares

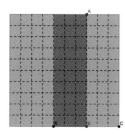

As the grid is 10×10, each dotted square is 1% of the large square

$$\therefore A = 30\% \text{ of } 144$$

$$= 43.2$$

OR as the length of folded rectangular area is the same as the length of the square and XB is $\frac{3}{10}$ the length of the square

$$A = \frac{3}{10} \text{ of } 144$$

$$= 43.2$$

25

Folded Area, A = 48 squares

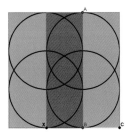

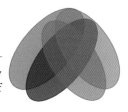

As XB is the radius of the circle, $3XB = 12$, $XB = 4$

$$A = 4 \times 12$$
$$= 48$$

Aside: The four circles do not quite overlap to allow each region to represent each and every combination of four sets as in the following Venn diagram.

26

Folded Area, A = 54 squares

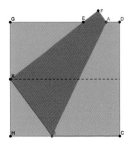

Let $XB = x$, then $HB = 12 - x$
Applying Pythagoras' Theorem on $\triangle HXB$,

$$x = \frac{15}{2}$$

As $HX = \frac{12}{2}$ and $HB = \frac{9}{2}$,

$\triangle BHX \sim \triangle 3,4,5$. scale factor $\frac{3}{2}$

$\triangle BHX \sim \triangle XGE \sim \triangle AFE$
As $XG = 6$, $GE = 8$ and $XE = 10$

$FE = 12-XE$ (*from the fold*) $= 2$
$\therefore$ As $FE = \frac{1}{2}$, $EA = \frac{5}{2}$ $FA = \frac{3}{2}$
$(GE + EA + AD = 12)$
As the folded area is a trapezium,

$$A = \frac{1}{2} \times 12 \times \left(\frac{15}{2} + \frac{3}{2} \right)$$

$$= 6 \times 9 = 54$$

OR $A = 144 - $ *Area AGHB*

$$= 144 - \frac{1}{2} \times 12 \times \left(\frac{9}{2} + \frac{21}{2} \right) = 54$$

By measurement $A = 0.5 \times 12 \times (7.5 + 1.5) = 54$

27

Folded area, A = 56 squares

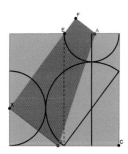

The radius of the smallest semicircle = 3. If the radius of the semicircle centred on X is r, then

$$3 + r + \sqrt{(3^2 + r^2)} = 12 \text{ using the radii of}$$

the largest semicircle

$$3^2 + r^2 = (9 - r)^2, \quad \therefore r = 4$$

ΔDEX has sides 6, 8, 10 so $EF = 2$ and from similar triangles

$$AF = \frac{2}{6} \times 8 = \frac{8}{3}; \ BX = \frac{4}{6} \times 10 = \frac{20}{3}$$

$$A = \frac{1}{2} \times 12 \times \left(\frac{20}{3} + \frac{8}{3}\right)$$

$$= 2 \times 28 = 56$$

By measurement,

$$A = \frac{1}{2} \times 12 \times (6.7 + 2.7) = 56.4$$

28

Folded Area, A = 36 squares

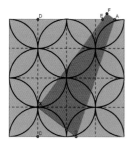

The radii of the circles is 3, so applying Pythagoras' Theorem on

ΔBXC and the similarity of triangles $\Delta BXC, \Delta XDE, \Delta AFE$

$BX = 5, AF = 1$

$$A = \frac{1}{2} \times 12 \times (5 + 1) = 36$$

Aside: The folded area is the same as in question 23. Is there a connection? Are there other points for X which create a folded area of 36?

29

Folded area, $A = 60\dfrac{12}{25} = 60.48 = 60.5$ squares (1 d. p.)

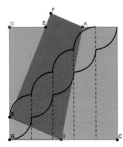

The semicircles define GX as $\dfrac{12}{5}$

Let GB = x,
then by applying Pythagoras' Theorem on $\triangle BXG$

$x = \dfrac{144}{25}$ *this is GX². Coincidence?*

$AX^2 = AF^2 + XF^2 = DX^2 + DA^2$

$\therefore$ *As $AF = (12 - DA)$, $AF = \dfrac{96}{25}$*

$A = \dfrac{1}{2} \times 12 \times \left(\dfrac{156}{25} + \dfrac{96}{25}\right)$

$= 6 \times 10\dfrac{2}{25} = 60\dfrac{12}{25}$

By measurement,
$A = \dfrac{1}{2} \times 12 \times (6.2 + 3.8) = 60$

30

Folded area, $A = 72\sqrt{3} - 72 \sim 52.707658\ldots\ldots = 52.7$ squares (1 d. p.)

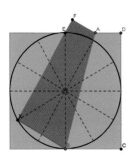

As XB forms an equilateral triangle with the centre of the circle $XB = 6$.

$\widehat{ABE} = 15°$. $\therefore AE = 12\tan(15°) = 24 - 12\sqrt{3}$

From the fold $AF = 6 - AE = 12\sqrt{3} - 18$

$A = \dfrac{1}{2} \times 12 \times (6 + 12\sqrt{3} - 18)$

$= 72\sqrt{3} - 72$

By measurement, $A = \dfrac{1}{2} \times 12 \times (6 + 2.8) = 52.8$ (1 d.p.)

31

Folded Area, A $= 360 - 216\sqrt{2} = 54.529870...... = 54.5$ *squares* (1 *d. p.*)

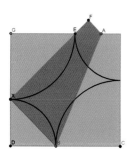

As XB = BC from the fold and
XD = DB = 12 – BC
Let XB = x, then

$$2(12 - x)^2 = x^2$$

$$\sqrt{2}(12 - x) = x$$

$$(\sqrt{2} + 1)x = 12\sqrt{2}$$

$$x = 12\sqrt{2}(\sqrt{2} - 1)$$

$$= 24 - 12\sqrt{2}$$

As GX = x, EX = $\sqrt{2}x$

$$AF = FE = 12 - \sqrt{2}x$$

$$A = \frac{1}{2} \times 12 \times (24 - 12\sqrt{2} + \{12-(24\sqrt{2} - 24)\})$$

$$= 360 - 216\sqrt{2}$$

By measurement, $A = \frac{1}{2} \times 12 \times (7 + 2.1) = 54.6$

32

Folded Area, A $= 144 - 72\sqrt{2} = 42.176623...... = 42.2$ *squares* (1 *d. p.*)

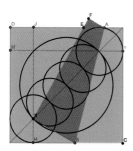

Let radius of small circles = r,
XB = x, FA = y

Along MN $r + \dfrac{4r}{\sqrt{2}} + r = 12$

$$\therefore r = 6\sqrt{2} - 6$$

From $\triangle XBH$ $x^2 = r^2 + (12 - r - x)^2$

$$\therefore x = \frac{1}{7}(78 - 30\sqrt{2})$$

From $\triangle XFA$ *and* $\triangle XJA,$

$$y^2 + 12^2 = (12 - r)^2 + (12 - r - y)^2$$

$$\therefore y = \frac{1}{7}(90 - 54\sqrt{2})$$

$$A = \frac{1}{2} \times 12 \times (x + y)$$

$$6 \times \frac{1}{7} \times (168 - 84\sqrt{2}) = 144 - 72\sqrt{2}$$

By measurement,

$$A = \frac{1}{2} \times 12 \times (5.1 + 1.9) = 42$$

$$= 42.0 \ (1 \ d.p.)$$

Aside: Which area is represented by $72\sqrt{2}$? Compare with the alternative answer to question 26. Can you show JA + HB = 12?

Folded Area, $A = 132 - 12\sqrt{30} \sim 66.273293\ldots\ldots = 66.3$ *squares* (1 *d. p.*)

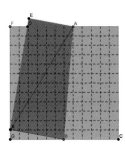

Let $FX = y$, $BX = x$ *Applying Pythagoras' Theorem*

$$AX^2 = EX^2 + EA^2 = FX^2 + FA^2$$

$$12^2 + 5^2 = y^2 + 7^2$$

$$y = \sqrt{120} = 2\sqrt{30} \; (= 10.954\ldots\ldots \text{ so not on the gridline!})$$

$$x = 17 - 2\sqrt{30} \; (= 6.045\ldots\ldots \text{ so not on the gridline!})$$

$$A = \frac{1}{2} \times 12 \times (17 - 2\sqrt{30} + 5)$$

$$= 132 - 12\sqrt{30}$$

By measurement, $A = \dfrac{1}{2} \times 12 \times (6 + 5) = 66 = 66.0$ (1 *d. p.*)

SECTION IV

Two folds

34

I Love Origami Dots

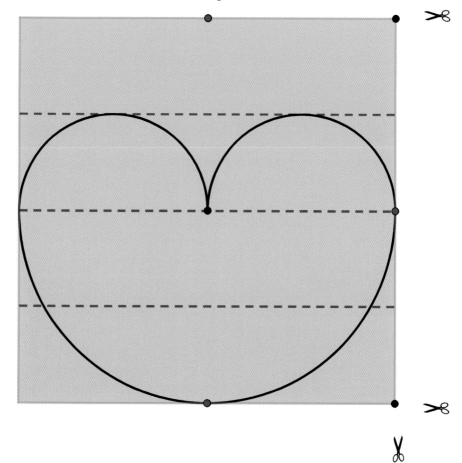

35

Rollers

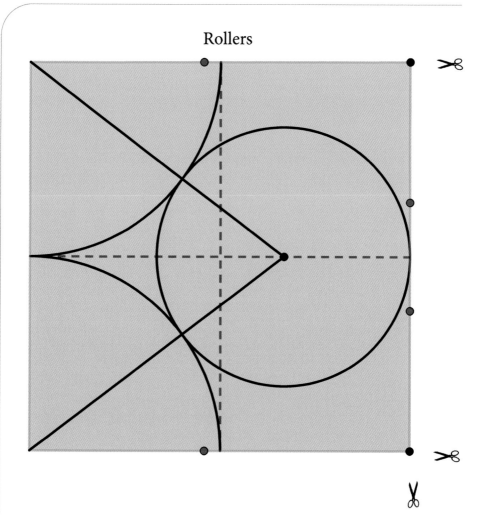

36

Yin-Yang Spiral

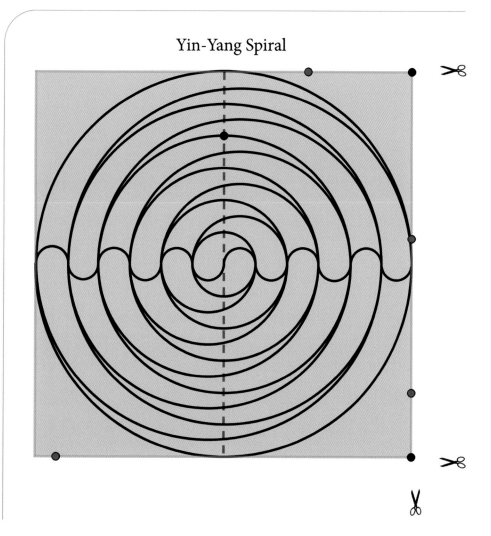

37

Arrowhead

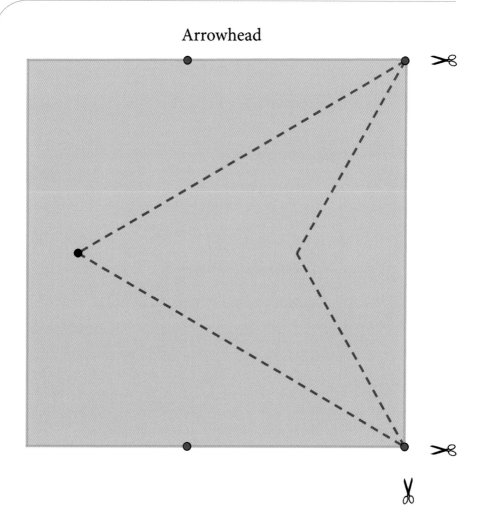

38

Ford Semicircles?

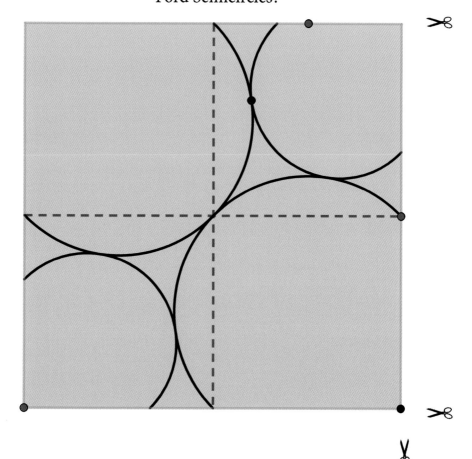

39

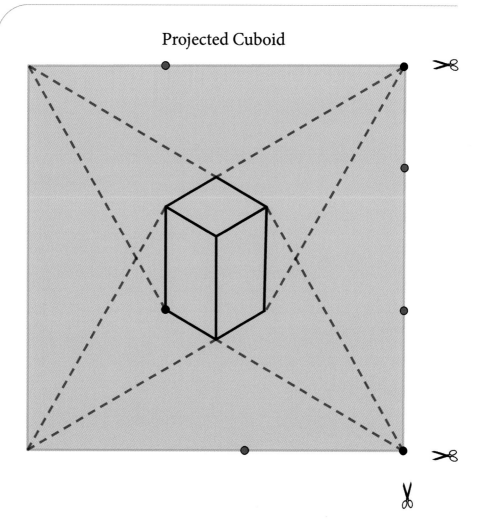

Projected Cuboid

40

Projected Pentagon

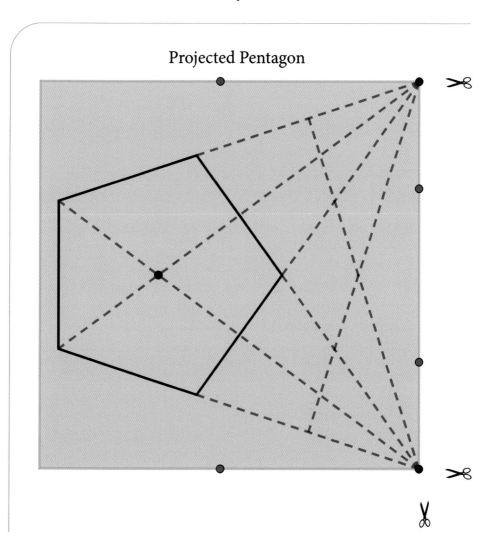

Solutions and Answers

34

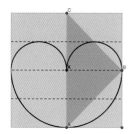

Folded area, A = 36 squares

The three semicircles define X as the centre of the square and XB is a line of symmetry of the folded triangle.

$$\therefore A = \frac{1}{2} \times 12 \times 6$$

$$= 36$$

35

Folded area, $A = 28\frac{1}{6} = 28.166666\ldots\ldots = 28.2$ squares (1 d.p.)

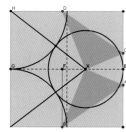

Let $XE = r$, $XA = x$

Applying Pythagoras' Theorem to $\triangle HGX$ and $\triangle AEX$

$$(r + 6)^2 = 6^2 + (12-r)^2$$

$$r = 4$$

$$(6-x)^2 + r^2 = x^2$$

$$x = \frac{13}{3} = 4\frac{1}{3}$$

$\triangle AEX \sim \triangle BFX$ so as $FB = \frac{6}{4} XE$,

$$XB = \frac{6}{4} XA = \frac{6}{4} \times \frac{13}{3} = \frac{13}{2}$$

The areas of $\triangle ABX$ and $\triangle CDX$ are equal because X lies on GE,

$$\therefore A = 2x = \frac{1}{2} \times \frac{13}{2} \times \frac{13}{3} = \frac{169}{6} \ 28\frac{1}{6}$$

By measurement

$2 \times 0.5 \times 6.5 \times 4.3 = 27.95 = 28.0$ (1 d.p.)

Aside: $\triangle XEC$ is similar to a right angled triangle with sides 5,12,13, the second primitive Pythagorean Triangle ($5^2 + 12^2 = 13^2$) after 3, 4, 5.

36

Folded area, A = 44 squares

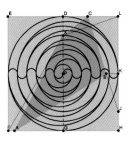

As the pattern forms a series of concentric circles around the centre of the square, P the given solution is a generalisation for any length of PX. Replace a with 4 for the set problem

Let PX = a, CL = x

In $\triangle DXC$ $(6 - a)^2 + (6-x)^2 = x^2$

$x = 6 - a + \dfrac{a^2}{12}$ As $\triangle LCJ \sim \triangle DXL$, $\left[\tan (\widehat{DXL}) = \dfrac{6}{6-a} \right]$ Note when X is at P then

$a = 0$, and $A = 36$ as in question 34

$LJ = \dfrac{6}{6-a} \left(6 - a + \dfrac{a^2}{12} \right) = 6 + \dfrac{a^2}{12 - 2a}$

$KJ = \dfrac{a^2}{12 - 2a}$ *and as* $\triangle BJK \sim \triangle CJL$

$BK = \dfrac{6-a}{6} \times \dfrac{a^2}{12 - 2a} = \dfrac{a^2}{12}$

Let AG = y. In $\triangle AGX$ $y^2 + (6 + a)^2 = (6 + y)^2$

$\therefore y = \left(a + \dfrac{a^2}{12} \right)$ *and AF* $= 6 - \left(a + \dfrac{a^2}{12} \right)$

Uncovered areas (bracketed values when $a = 4$)

$AFX = \dfrac{1}{2} \left(6 - a - \dfrac{a^2}{12} \right)(6 + a) = 18 - \dfrac{3a^2}{4} - \dfrac{a^3}{24}$ $\left(\dfrac{10}{3} \right)$

$DEFX = \dfrac{1}{2} \times 6 \,(12 + 6 - a) = 54 - 3a$ $\qquad$ (42)

$CDX = \dfrac{1}{2} (6 - a) \left(a - \dfrac{a^2}{12} \right) = 3a + \dfrac{3a^2}{4} + \dfrac{a^3}{24}$ $\left(\dfrac{8}{3} \right)$

$BLKC = \dfrac{1}{2} \times 6 \left(6 - a + \dfrac{a^2}{12} + \dfrac{a^2}{12} \right) = 18 - 3a + \dfrac{a^2}{2}$ $\quad$ (14)

$AHKB = \dfrac{1}{2} \times 6 \left(6 + a + \dfrac{a^2}{12} + \dfrac{a^2}{12} \right) = 18 + 3a + \dfrac{a^2}{2}$ $\quad$ (38)

Total area $= \left(108 - \dfrac{a^2}{2} \right)$,

$\therefore A = 144 - \left(108 - \dfrac{a^2}{2} \right) = 36 + \dfrac{a^2}{2}$ $\quad$ (44)

Note when X is at P then $a = 0$, and $A = 36$ as in question 34

OR to avoid this algebra, the same result could be obtained by calculating the folded area associated with each concentric circle (36.5, 38, 40.5, 44, 48.5 and 54) and working out the nth term of the resulting quadratic sequence.

By measurement

A = area $\triangle XBC$ + area $\triangle XAB$

= $0.5 \times 6.1 \times 3.3 + 0.5 \times 6.1 \times 11.1$

= $43.92 = 43.9$ (1 *d. p.*)

37

Folded area, $A = 24\sqrt{3} = 41.569219\ldots\ldots = 41.6$ *squares* (1 *d. p.*)

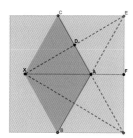

All the triangles such as $\triangle XCD$ are $1/2$ of an equilateral triangle side $4\sqrt{3}$

by moving $\triangle XCD$ *onto* $\triangle EAD$ *the folded area becomes identical to the area in Question 9.*

By measurement $A = 0.5 \times 12 \times 6.9 = 41.4$

Aside: What is the area of the arrowhead $AEXG$?

38

Folded area, $A = 45$ *squares*

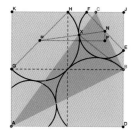

To first show that $\triangle ABC$ is a right angled triangle.

As M is the midpoint of GH it has co-ordinate (3,9) if A is (0,0).

Let N be (x, x) *from symmetry. As* $\triangle EJF$ *is isosceles,*

applying Pythagoras' Theorem on $\triangle MPN$

$(x - 3)^2 + (x - 9)^2 = (3\sqrt{2} + (12 - x)\sqrt{2})^2$

$x = 10$

As $MX:XN = 3:2$

$$X = \left(3 + \frac{3}{5} \times 7, 9 + \frac{3}{5} \times 1\right) = (7.2, 9.6)$$

Let CX = y = CJ, then $2.4^2 + (4.8 - y)^2 = y^2$

$\therefore y = 3$ *and C is* (9, 12)

$\therefore$ *As* $\dfrac{7.2}{9.6} = \dfrac{9}{12}$ *points A, X, C are colinear (they are all*

on the same line) and $\triangle CJB \sim \triangle BDA$, $\widehat{CBA} = 90°$.

$\therefore A = \dfrac{1}{2} CB \times AB = \dfrac{1}{2} \times \sqrt{6^2 + 3^2} \times \sqrt{12^2 + 6^2} = 45$

By measurement

$A = 0.5 \times 6.7 \times 13.4 = 44.89 = 44.9$ (1 d. p.)

Aside: Why the title Ford Semicircles? Ford circles have radii that link to fractions in the Farey Sequence where a fraction inbetween two others, say $\dfrac{a}{b}$ and $\dfrac{c}{d}$ is given by $\dfrac{a+c}{b+d}$. In the diagram if KJ = 1 then

$KH = \dfrac{1}{2}$, $KF = \dfrac{2}{3}$ and the horizontal distance of X from

the left is $\dfrac{3}{5}$.

39

Folded area, $A = 90 - 30\sqrt{3} = 38.038475\ldots\ldots = 38.0$ *squares* (1 d.p.)

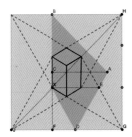

Let XF = x, FG = 12 – x as X is on HJ by symmetry,
   and JF = XF

As $\widehat{FGX} = 30°$ $XG = 2x$

$\therefore x^2 + (12 - x)^2 = (2x)^2$

$x^2 + 12x - 72 = 0$

$(x + 6)^2 = 108$

$x = 6\sqrt{3} - 6$

$FD = \dfrac{XF}{\sqrt{3}} = 6 - 2\sqrt{3}$

$\therefore XD = 12 - 4\sqrt{3}$

$A = area\ of\ (ABC + AEXC + EDX)$

As X is on HJ $\widehat{ABC} \therefore = 45°$ $\therefore$ *area of ABC* $= 1/2 \times 6 \times 6 = 18$

Area of AEXC $= \dfrac{1}{2}(6 - (6\sqrt{3} - 6))(6 + (12 + 4\sqrt{3})) = 144 - 78\sqrt{3}$

$$\text{Area of } EDX = \frac{1}{2}(12 - 6\sqrt{3})(12 - 4\sqrt{3}) = 48\sqrt{3} - 72$$

$$A = 90 - 30\sqrt{3}$$

By measurement $A = 0.5 \times 6 \times 6 + 0.5 \times 1.6 \times (6 + 5.1) + 0.5(4.4 \times 5.1) = 38.1$

40

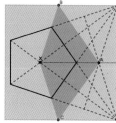

Folded area, $A = 36\sqrt{2 - \dfrac{2}{\sqrt{5}}} = 37.852640\ldots\ldots = 37.98 \ squares \ (1 \ d. \ p.)$

$$A = \frac{1}{2} \times 12 \times AX$$

$$AX = XE - AE$$

By considering $\triangle EXD$ and $\triangle EAD$

$$AX = 6 \tan(54°) - 6 \tan(18°)$$

$$A = 6 \times 6 \ (\tan(54°) - \tan(18°))$$

$$= 36\left(\sqrt{1 + \frac{2}{\sqrt{5}}} - \sqrt{1 - \frac{2}{\sqrt{5}}}\right)$$

$$= 36\left(\sqrt{2 - \frac{2}{\sqrt{5}}}\right)$$

By measurement, $A = 0.5 \times 12 \times 6.3 = 37.8$.

The surd form of the answer was found using WolframAlpha®, an online computer algebra system. This demonstrates the power of working with a computer or calculator to support difficult and/or long computations. This is especially true if the resulting representation leads to other insights about the problem. Is that the case here?

Compare the answer with the answer in question 11. What is the same in both problems? What is different?

Afterword

Well done, having made it to the end of the book you have already shown the essential mathematical skills of resilience and curiosity. However, the challenge does not end here. Most of the puzzles have alternative points that the corners can be folded to and other intriguing questions come to mind.

Why does the intersection of the creases in the two-fold puzzles always seem to lie on the halfway line of the square?

Or can any point in the square be defined using the construction rules of Origami Dots?

The most important questions are of course yours. See what you can discover. Happy searching.   AP